< GET CODING! >

Get Coding with Raspberry Pi™

By Jenna Vale

Gareth Stevens
PUBLISHING

Please visit our website, www.garethstevens.com. For a free color catalog of all our high-quality books, call toll free 1-800-542-2595 or fax 1-877-542-2596.

Cataloging-in-Publication Data
Names: Vale, Jenna.
Title: Get coding with Raspberry Pi / Jenna Vale.
Description: New York : Gareth Stevens Publishing, 2024. | Series: Get coding! | Includes glossary and index.
Identifiers: ISBN 9781538288665 (pbk.) | ISBN 9781538288672 (library bound) | ISBN 9781538288689 (ebook)
Subjects: LCSH: Raspberry Pi (Computer)–Juvenile literature. | Python (Computer program language)–Juvenile literature. | Microcomputers–Programming–Juvenile literature. | Computer programming–Juvenile literature.
Classification: LCC QA76.8.R15 V35 2024 | DDC 004.1675–dc23

Published in 2024 by
Gareth Stevens Publishing
2544 Clinton Street
Buffalo, NY 14224

Portions of this work were originally authored by Patricia Harris and published as *Understanding Coding with Raspberry Pi*. All new material in this edition authored by Jenna Vale.

Designer: Leslie Taylor
Editor: Megan Kellerman

Photo credits: Cover goodcat/Shutterstock.com, (back illustrations) StonePictures/Shuttertstock.com; cover & series art (coding background) BEST-BACKGROUNDS/Shutterstock.com; p. 5 Argonne National Laboratory/https://commons.wikimedia.org/wiki/File:AVIDAC_-_First_Argonne_Computer_(1953).jpg, (inset) trickyaamir/Shutterstock.com; p. 7 goodcat/Shutterstock.com; p. 9 mattcabb/Shutterstock.com; p. 11 Andy Oakley/https://commons.wikimedia.org/wiki/File:Cjam-html-server-tutorial-terminal.png; p. 15 Gareth Halfacree/https://commons.wikimedia.org/wiki/File:Raspberry_Pi_Zero_TRNG_Project_(22950121039).png; p. 17 Dj Walker-Morgan/Flickr.com; p. 19 Raspberry Pi Foundation/https://commons.wikimedia.org/wiki/File:RPi_learning_Physical_Computing_with_Python.png; pp. 21 (geany logo) unknown author/https://commons.wikimedia.org/wiki/File:Geany_logo.svg, (screenshot) Path slopu/https://commons.wikimedia.org/wiki/File:Geany-example_C%2B%2B_program_with_output.png; p. 25 (top) wavebreakmedia/Shutterstock.com, (Scratch screenshot) Scratch is developed by the Lifelong Kindergarten Group at the MIT Media Lab; p. 27 Bogdan Vija/Shutterstock.com; p. 29 Daniel Chetroni/Shutterstock.com.

Printed in the United States of America

CPSIA compliance information: Batch #CS24GS: For further information contact Gareth Stevens at 1-800-542-2595.

CONTENTS

Words in the glossary appear in **bold** the first time they are used in the text.

In the Palm of Your Hand

Computers have many parts and can do a lot of different things. They used to be so big, they would fill entire rooms! It would have been hard for people in the 1950s to believe computers would be as small and widely used as they are today.

In 2012, a charity called the Raspberry Pi Foundation created a new computer—and it was only the size of a credit card! It was the Raspberry Pi, and the foundation wanted to use it to help students learn not only how to use computer programs, but how to make them.

This is a computer from the year 1953. Today, a computer such as the Raspberry Pi fits in the palm of your hand.

Just the Computer

The Raspberry Pi is all computer—it doesn't even come inside a case! Users have created cases using many different things: a popular candy tin that's just the right size, LEGO blocks, and even 3D printers. This DIY (do it yourself) way of thinking helps users have fun while learning about computers and coding.

On the Board

Raspberry Pi doesn't come in a case, so you can see all the hardware, or components, of the computer. When you first open a new Raspberry Pi, you'll find a green **circuit board**. This board has most of the processing devices a regular computer has.

The Raspberry Pi comes with ports needed to connect other devices to it, including a monitor, keyboard, mouse, power cord, headphones, and other components. Many devices connect to the Raspberry Pi using audio and video cables, as well as common **USB cords**. The Raspberry Pi also has a port for a memory card.

This is one model of the Raspberry Pi. The Raspberry Pi Foundation has created many different models.

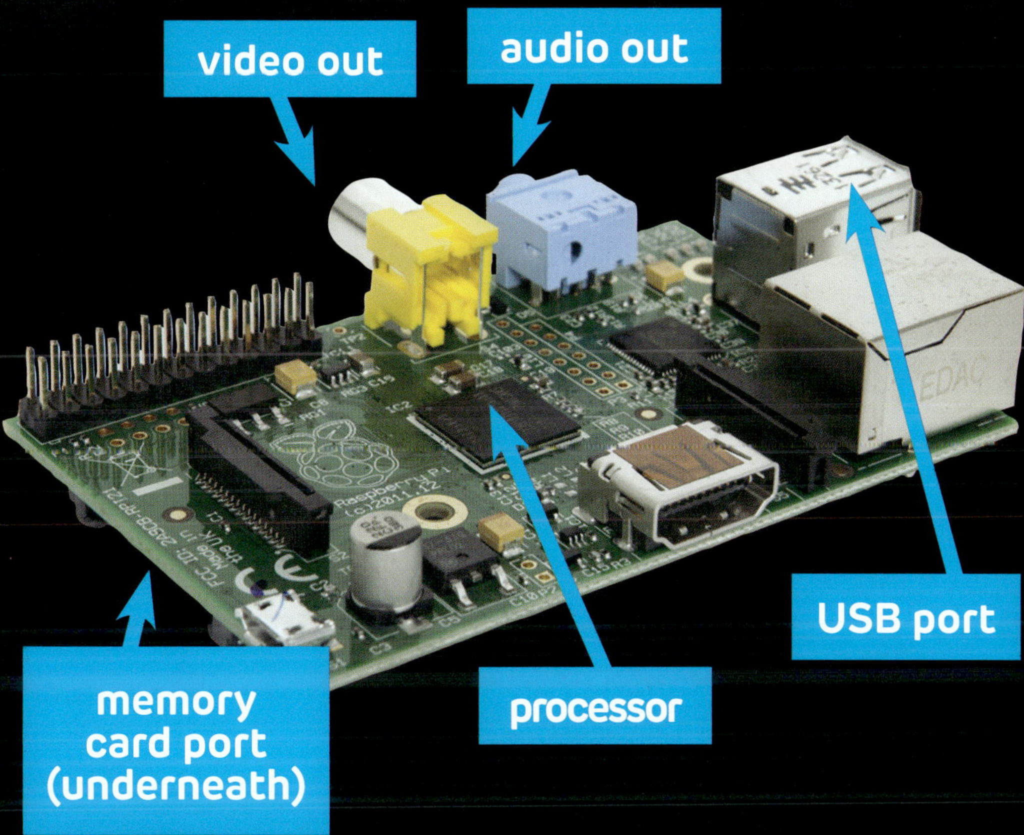

SBC

The Raspberry Pi is a single-board computer, or SBC. An SBC is a whole computer on just one circuit board! This means an SBC can do most things that a regular computer can do. SBCs are also very compatible with other types of devices, meaning they can be connected to many things.

Starting Up the Pi

The programs or applications that run on a computer are called software. This includes the computer's operating system (OS). An OS manages all the other programs and controls the computer's operations. The Pi runs on the Raspberry Pi OS, a special edition of the Linux OS.

When you turn on the Pi, you will see the **graphical** desktop on your monitor. This shows icons for the applications. One important icon in the desktop drop-down menu is the shutdown icon. Make sure you shut the Pi down before unplugging it, or else you could damage the SD card, which is the storage device that holds the OS.

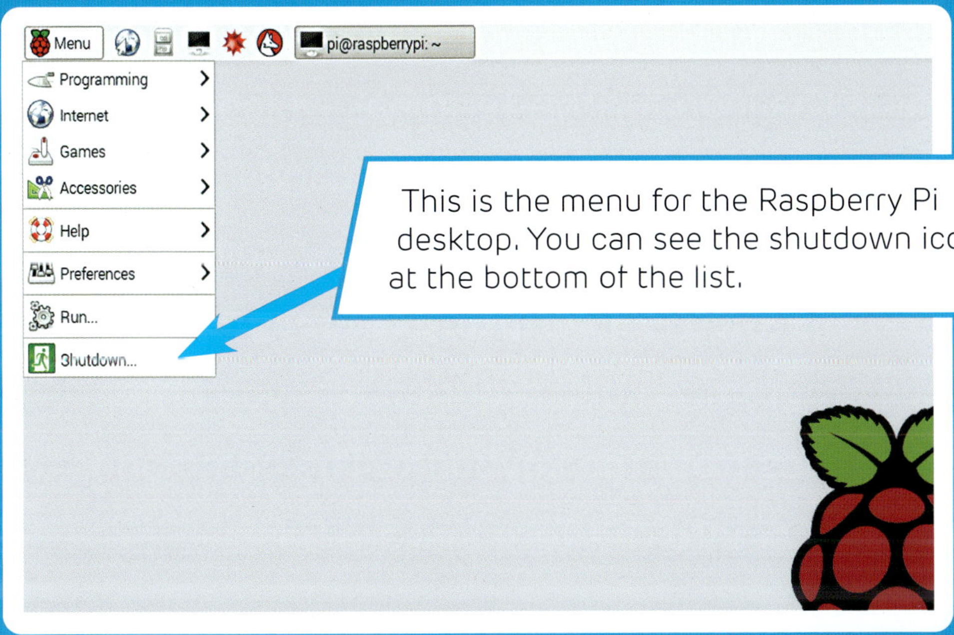

This is the menu for the Raspberry Pi desktop. You can see the shutdown icon at the bottom of the list.

Linux and UNIX

Raspberry Pi OS and Linux are open-source, or free, software. Finnish coder Linus Torvalds created Linux as an improvement over the open-source OS called UNIX. Torvalds created what's known as a kernel, which is the core part of an OS. The kernel helps manage the memory of other programs and helps them open quickly.

The Terminal

The bar at the top of the desktop display has an icon that looks like a computer screen. This is the icon for the Terminal, which is a window you need to run commands. The Terminal lets you use the Bourne-Again SHell, or Bash, which is a **scripting language** and a shell program. A shell program lets you enter commands from the keyboard. The commands you'll enter into the Terminal are Bash commands.

When you open the Terminal, you will be in your home **directory**. You will see the prompt "pi@raspberrypi: ~ $". All your commands are entered after the "$" character.

The first "pi" in "pi@raspberrypi" is the **username** the Pi comes with. If you change the username to something else, then "pi" will be exchanged with your username.

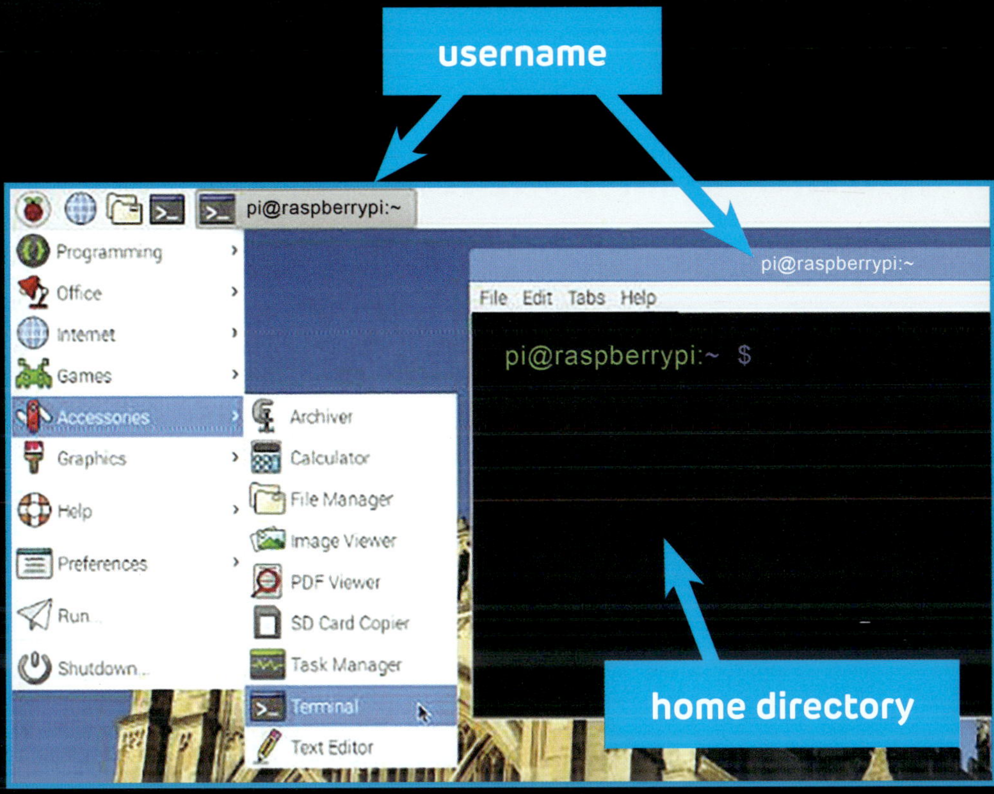

Going Beyond with Bash

Learning to write commands on the Raspberry Pi can help you with coding on other Linux based devices later. This is because Bash is used on any other Linux-based device. The short scripts you can write with Bash to run programs will set you on your way to writing programs!

Learning the Commands

Using commands in the Terminal is very useful. The Terminal can even be used to control the Raspberry Pi so you don't have to click around with your mouse! You can use the Terminal to look around at what files the Pi already has or work with any files you've added to the computer.

You can also use the Terminal to run programs and applications. For example, if you want to run the application called Raspberry Pi **Configuration**, you would type "rasp-config" into the Terminal. See the chart on the next page for commonly used commands and some examples.

These are some common Bash commands you'll use in the Terminal. Add the directory, file name, or text you want to print after the command.

Basic Bash Commands

COMMAND	WHAT IT DOES	EXAMPLE COMMAND
ls	lists the files in the current directory (That's a lowercase "L", not an "i" or "1"!)	**$ ls -l**
cd	changes the current directory	**$ cd Desktop**
echo	makes the program display text, or words	**$ echo Hello World**
touch	will bring the existing file's date and time data up to date OR will create a new file under that name if the file doesn't already exist	**$ touch Test**
rm	removes a file (This is **permanent**, so be sure you want to remove the file for good!)	**$ rm Test**
--help	asks for help about any command	**$ rm --help**

The Case Matters

Commands in the Terminal are case-sensitive. This means the result of a command depends on whether or not the command had uppercase or lowercase letters. For example, if you type the command "cd desktop", you might get an error message because the "d" in "desktop" needs to be uppercase for the command to work.

Input and Output Pins

The keyboard and mouse let you add input to the computer, but what about output? Computers have different ways to show output, such as a monitor and speakers. Raspberry Pi works the same way, but these other components aren't included with it.

The Raspberry Pi has USB ports and a port to attach it to a TV. It also has a set of general-purpose input/output pins (GPIO) to make connections. These pins are wired to the board, but don't have special jobs like USB ports. That's why they're called general purpose. These pins allow users to do things that would be harder to do on a regular computer.

There is a different voltage, or amount of electrical power, for some GPIO pins. Be very careful! Make sure you know which pin is which before making any connections.

Map It!

If you'd like to learn more about the GPIO pins on your Raspberry Pi, you can ask the Terminal for help. Type the command "pinout" and the Terminal will show you a labeled map of the Pi. It will also tell you more about the hardware, ports, and model of your Pi.

The GPIO pins can be controlled using commands in the Terminal. You could connect an **LED** light to the GPIO pins and write the code to turn the light on and off. To program the pins on Raspberry Pi, you must be a "superuser" and not just the "pi user." This means you will be able to do more than a regular user. In the Terminal, type in "sudo su". You will see that the prompt changes to "root@raspberrypi:/home/pi#". The command "exit" takes you back.

To turn on the LED light, you'll use the "echo" command. The command "echo 1 > value" turns the light on. The command "echo 0 > value" turns the light off.

With Great Power...

It's important to be careful when acting as a superuser! You could damage the OS because you are much more powerful as a superuser than as a regular user. If you make a mistake as a superuser, you could hurt the data on the SD card that contains the OS.

The "echo" command works for both displaying text in your program and turning LED lights on and off.

Physical Computing

The projects you do with the Pi's GPIO pins are examples of **physical** computing. This means that what you program into the computer affects the real world. Physical computing involves programming different **sensors** to take in data (input) and then have an effect on the real world (output).

A more advanced LED project is programming the Pi to control miniature, or tiny, stoplights. You'll need a few extra items for this project, including a breadboard, which is an extra board used for testing electrical circuits. Once the Pi is set up, you can write the code using the Python **programming language**, which you'll learn about in the next chapter!

The miniature stoplights project will teach you about programming a button (human input), a buzzer (sound output), and LEDs (light output).

The Internet of Things

Many devices we interact with daily are now smart devices, which means they can send and receive data using the internet in some way. This is called the Internet of Things, and it is an advancement of physical computing. There are smart forms of so many things now, including speakers, refrigerators, and even washing machines!

Coding with Python

The Raspberry Pi comes with the Python programming language. It also comes with an integrated development environment (IDE), or editor, called Geany. The IDE allows you to write Python code and see it displayed in one application. This means you don't have to write code in an editor and then go to the Terminal window to see the results.

Geany has a Run menu choice that lets you run your program after it's saved. Choosing Run will open the script in a new window. You can also use the Terminal tab at the bottom of the Geany editor to run the script from there directly.

This is the Geany editor. Python programs are saved with a file name that ends in the ".py" file extension.

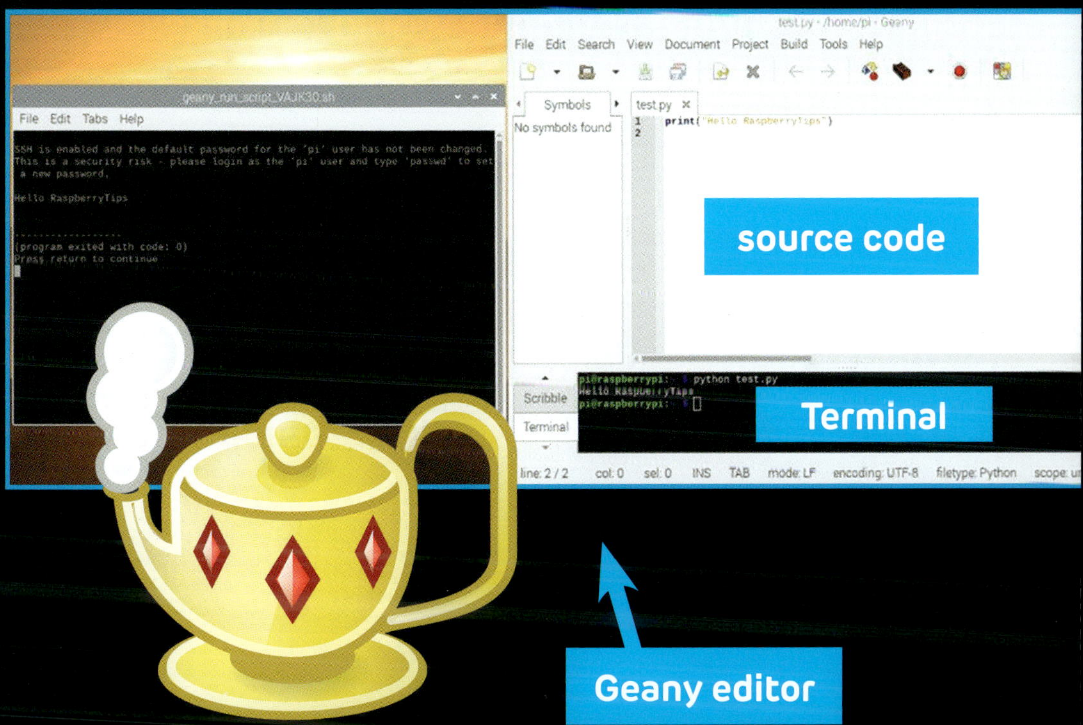

Working with Geany

Geany is a very useful IDE. It can run on devices that use macOS, Microsoft, or the Linux operating system. You can also use it with more than 50 programming languages! This makes Geany a very helpful IDE to learn coding with. Besides Python, supported languages include JavaScript, HTML, Ruby, and C++.

for the user to type on. When the user enters their name, the program will greet them by name. This is what the output for the program will look like:

```
What is your name?
Jessie
Hello Jessie
```

Making Loops

A loop is a set of commands grouped together so they repeat. Loops make coding easier and quicker, because you don't have to retype the code every time you want it to repeat. In Python, a "while" loop is used to tell the loop to repeat while a certain event is happening.

Remember the stoplights project from page 18? Here is what the first step of coding each component looks like!

Button

```
1   from gipozero import Button
2   button = Button(21)
3   while True:
4       button.wait_for_press()
5       print("Pressed")
6       button.wait_for_release()
7       print("Released")
```

1. Tells the program we're going to be using a button.
2. Identifies the button as GPIO pin #21.

3–7. Creates a loop to test what happens while the button is being pressed (program prints the word "Pressed") and released, or let go of (program prints the word "Released").

LED Lights

```
1   from gipozero import Button, LED
2   button = Button(21)
3   led = LED(25)
4   while True:
5       button.wait_for_press()
6       led.on()
7       button.wait_for_release()
8       led.off()
```

1. Adds the LED light to the program.
2. Same code as before.
3. Identifies the LED as GPIO pin #25.

4–8. Changes the loop so the LED light turns on when the button is pressed and turns off when the button is released.

Buzzer

```
1   from gipozero import Button, LED, Buzzer
2   button = Button(21)
3   led = LED(25)
4   buzzer = Buzzer(15)
5   while True:
6       button.wait_for_press()
7       led.on()
8       buzzer.on()
9       button.wait_for_release()
10      led.off()
11      buzzer.off()
```

1. Adds the buzzer to the program.

2–3. Same code as before.

4. Identifies the buzzer as GPIO pin #15.

5–11. Changes the loop so the buzzer makes a sound under the same conditions that turn on the LED.

Drag and Drop Coding

The Pi also comes with a programming language called Scratch. It's found under Programming in the drop-down menu on the desktop. The program opens in its own window.

Instead of using words to make commands, Scratch uses colorful blocks that you drag and drop into an on-screen workspace. Scratch uses different types of blocks for movements, sounds, events, and more. When users put the blocks together, they create **animations**. Some blocks have curved tops and sit at the start of a program. Others have cutouts at the top and tabs at the bottom because they are meant to link together or stack.

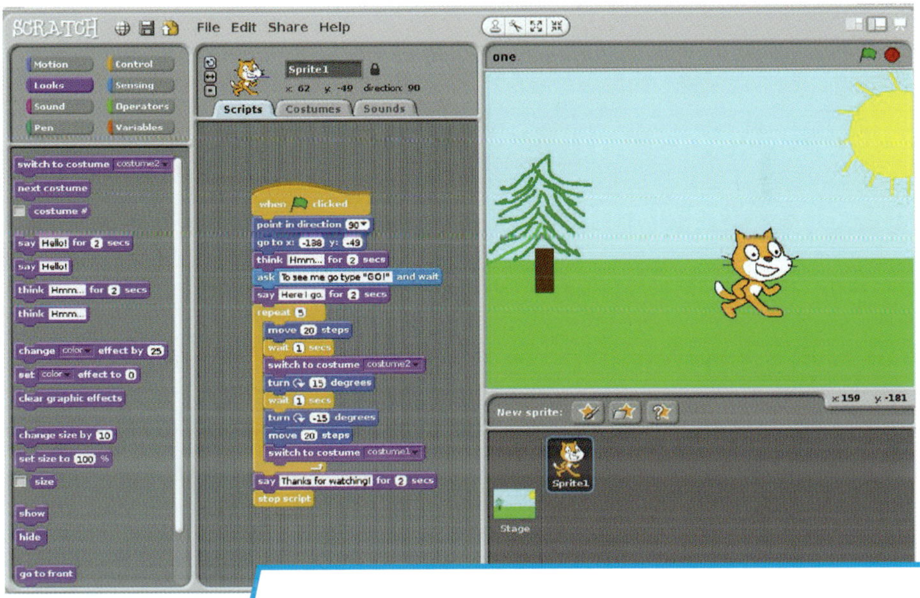

This is what Scratch looks like. It uses a graphical user interface, or GUI, which means you code using pictures instead of words.

Scratch Blocks

You'll learn the same coding principles in Scratch that you've explored in Python—just with graphics instead of words! You can even make loops in Scratch using Control blocks. You drag a series of blocks into the Control block to make a loop. This way, a series of commands can act as one block.

To start coding with Scratch, open the program from the drop-down menu on the desktop. If the program Scratch 3 isn't already included, you can type the command "sudo apt install scratch3" into the Terminal to install the program.

As with Python, you can do many coding projects with Scratch on the Pi. You can code games or program a robot. If you're switching from Python, you can even do the stoplights project. At the bottom of the Blocks menu, click Add Extension and choose Raspberry Pi Simple Electronics. This will give you the set of coding blocks needed to do electrical projects with the Pi.

You can practice coding with Python or Scratch on the Raspberry Pi right away.

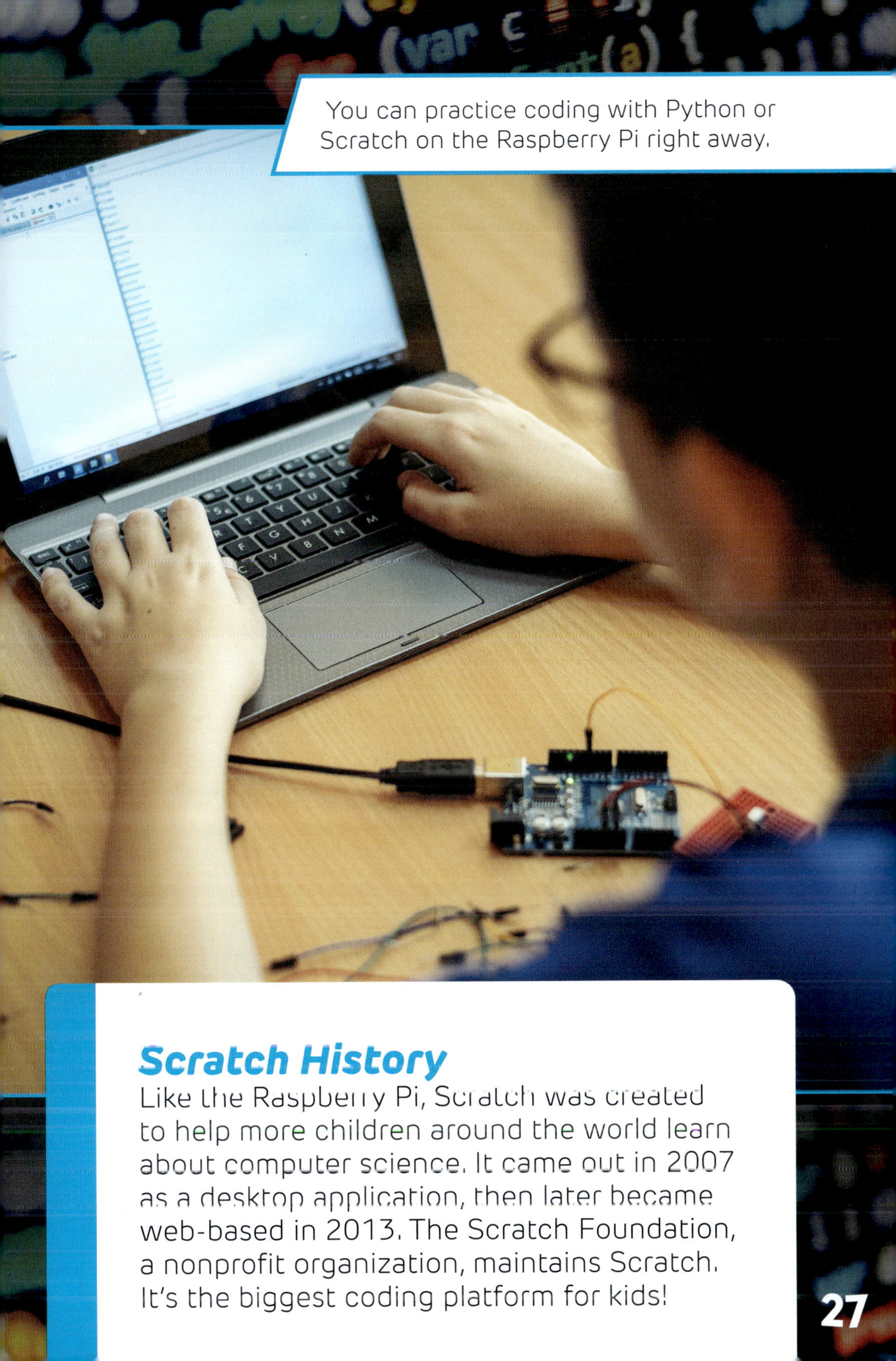

Scratch History

Like the Raspberry Pi, Scratch was created to help more children around the world learn about computer science. It came out in 2007 as a desktop application, then later became web-based in 2013. The Scratch Foundation, a nonprofit organization, maintains Scratch. It's the biggest coding platform for kids!

The Whole Pi

Working with the Raspberry Pi can teach you so many things: coding, physical computing, making games, and so much more. Remember the Internet of Things? You can even create your own smart home system with a Raspberry Pi!

The Raspberry Pi Foundation set out to make it easier for kids to learn about computers, and today they have helped millions of children around the world do just that. You can code games with Scratch, write programs with Python, or experiment with physical computing. Whatever you're interested in learning, you can learn how to do it with the Raspberry Pi.

There are many choices for getting started with a Raspberry Pi. Which model would you like to learn with?

Big Dream, Small Package

A computer scientist named Eben Upton created the Raspberry Pi. At first, he thought that Python would be the perfect—and only—programming language that would work for the Pi. The "Pi" in the name is even short for "Python." However, Upton and the Raspberry Pi team were able to create a computer that could do so much more than they dreamed!

GLOSSARY

animation: Drawings or computer images that appear to move.

configuration: The arrangement of certain elements, such as computer settings.

circuit board: A thin, rigid board that holds the electrical parts of a device such as a computer.

directory: A computer "folder" that makes finding files easier. It can also store other directories.

graphical: Having to do with graphics, or pictures.

permanent: Unable to be changed.

physical: Having to do with the material world.

programming language: A language made to give instructions to a computer.

LED: A light-emitting diode. Also, an indicator light on an electronic device.

scripting language: A programming language specifically made for the creation of short programs to perform simple tasks.

sensor: A device that reacts to a change in the world around it and then performs a task.

USB cord: A wire with very common connectors, used to transfer power and files from one device to another.

username: A special name that identifies who is using a program, application, or computer.

FOR MORE INFORMATION

Books

Briggs, Jason R. *Python for Kids, 2nd Edition: A Playful Introduction to Programming*. New York, NY: No Starch Press, 2022.

Halfacree, Gareth. *The Official Raspberry Pi Beginner's Guide: How to Use Your New Computer*. Cambridge, UK: Raspberry Pi Trading Ltd, 2020.

McCue, Camille. *Getting Started with Coding: Get Creative with Code!* Hoboken, NJ: Wiley, 2020.

Websites

Python
www.python.org
This is the official website for the Python programming language.

Raspberry Pi Foundation
www.raspberrypi.org
This is the official website for all models of the Raspberry Pi.

Scratch
www.scratch.mit.edu
This is the official website for the Scratch programming language.

Publisher's note to educators and parents: *Our editors have carefully reviewed these websites to ensure that they are suitable for students. Many websites change frequently, however, and we cannot guarantee that a site's future contents will continue to meet our high standards of quality and educational value. Be advised that students should be closely supervised whenever they access the internet.*

INDEX